Deciduous Forests

Jennifer Hurtig

WEIGL PUBLISHERS INC.

Published by Weigl Publishers Inc.
350 5th Avenue, Suite 3304, PMB 6G
New York, NY 10118-0069
USA

Web site: www.weigl.com

Library of Congress Cataloging-in-Publication Data

Hurtig, Jennifer.
    Deciduous forests / Jennifer Hurtig.
    p. cm. — (Biomes)
    Includes index.
    ISBN 1-59036-440-6 (hard cover : alk. paper) —
    ISBN 1-59036-441-4 (soft cover : alk. paper)
    1. Forest ecology—Juvenile literature. I. Title. II. Biomes (Weigl Publishers)
QH541.5.F6H87 2006   577.3—dc22   2006001031

Printed in China
1 2 3 4 5 6 7 8 9 0 10 09 08 07 06

Project Coordinator
Heather Kissock

Designers Warren Clark,
Janine Vangool

Cover description: Maple trees
are known for their tasty sap.
This sap can be turned into
maple syrup and maple sugar.

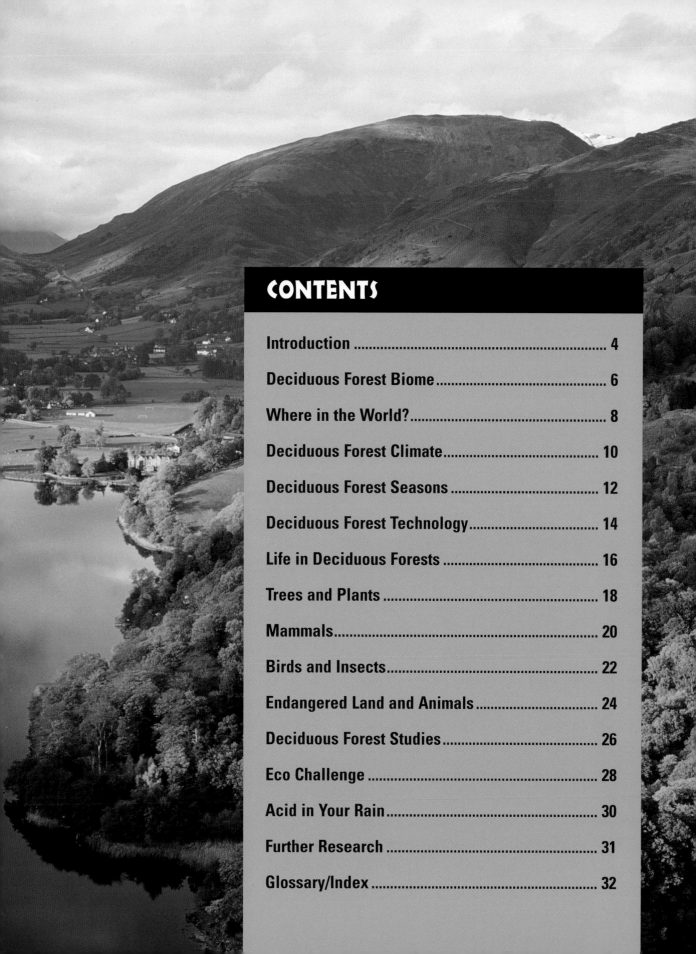

CONTENTS

Introduction

Earth is home to millions of different **organisms**, all of which have specific survival needs. These organisms rely on their environment, or the place where they live, for their survival. All plants and animals have relationships with their environment. They interact with the environment itself, as well as the other plants and animals within the environment. This interaction creates an **ecosystem**.

Different organisms have different needs. Not every animal can survive in extreme climates. Not all plants require the same amount of water. Earth is composed of many types of environments, each of which provides organisms with the living conditions they need to survive. Organisms with similar environmental needs form communities in areas that meet these needs. These areas are called biomes. A biome can have several ecosystems.

The trees of the New England states, including Connecticut, are known for their autumn colors.

The maple tree is deciduous. It loses its leaves every fall.

The word "deciduous" comes from the Latin word *deciduus*, which means "to fall off." Deciduous means to fall off or shed at a certain season. Deciduous forests contain trees that lose their leaves in the autumn and winter months. Then, the leaves grow back in spring and summer. Deciduous forests may also contain some plants that do not lose their leaves.

Deciduous forests have tall, broad-leafed hardwood trees. Small flowers, shrubs, ferns, and grasses grow there as well. Deciduous forests are found across North America, Europe, and parts of China and Japan. Most of these forests are found near an ocean. Deciduous forests are found in places where the climate is moist and mild. The climate is usually characterized by four distinct seasons.

FASCINATING FACTS

The United States and Europe have turned some of their remaining deciduous forests into national parks and nature preserves.

Open areas in deciduous forests often have grass and streams that provide homes for newts, frogs, and fish.

Deciduous Forest Biome

T he first deciduous trees developed about 180 million years ago when the temperature on Earth was warmer than it is today. North America and northern Europe were covered with walnut, oak, maple, hickory, and chestnut trees. Then the ice ages arrived, and the movement of glaciers destroyed many of the forests. Ten thousand years ago, after the Ice Age, birch trees were some of the first trees to grow back.

Other deciduous tree species also reappeared after the ice ages. These trees were able to grow back due to the way they reproduce. Grown trees release seeds. These seeds are often moved to other areas by the wind or insects. If this new area is fertile, the seeds will sprout and grow into trees. Following the Ice Age, seeds that had been buried under ice began to thaw and germinate, allowing for the regrowth of deciduous forests.

The leaves of the shagbark hickory tree turn a rich gold when autumn arrives.

Deciduous forests are sometimes called temperate deciduous forests. This is because they often grow in the middle latitudes around the globe. This part of Earth is called the temperate zone. The northern temperate zone lies between the Tropic of Cancer and the Arctic Circle. Between the Tropic of Capricorn and the Antarctic Circle lies the southern temperate zone. In temperate zones, summers and winters are of equal length.

Coniferous forests, with their many evergreen trees, often border deciduous forests. Some evergreens are even found in deciduous forests. Deciduous forests are closer to the equator than many coniferous forests. They also have longer growing seasons.

FASCINATING FACTS

Forests in North America's New England and Great Lakes areas are fairly young because they were renewed after the Ice Age. Older forests in the southern Appalachian region have more tree species.

Humans have a huge impact on the deciduous forest biome. People have resided in or near deciduous forests for thousands of years.

WHERE IN THE WORLD?

Most deciduous forests are found in the Northern Hemisphere, although some are found in the Southern Hemisphere. This map shows where the world's major deciduous forests are located. Do you live near a deciduous forest? Where is the deciduous forest closest to you?

Deciduous Forests

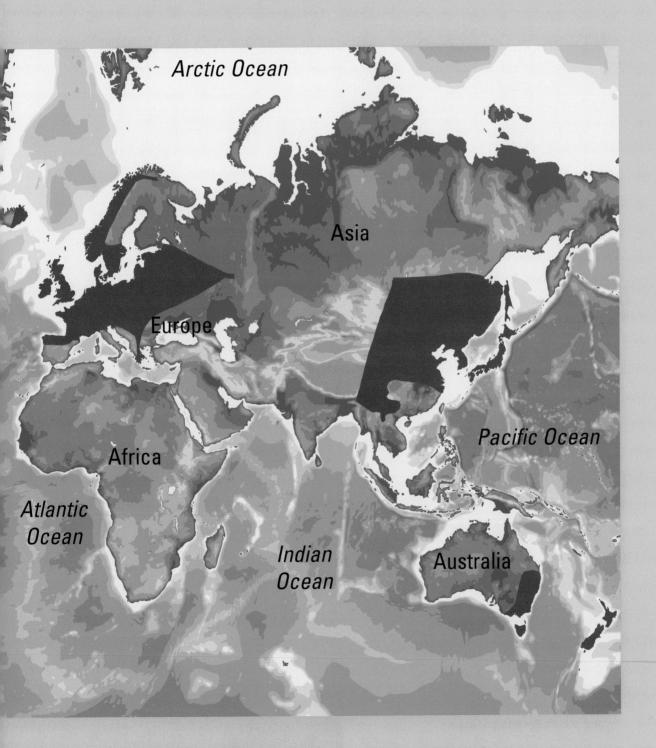

Deciduous Forest Climate

T he deciduous forest biome lies between the polar regions and the tropics. Due to its location, this biome receives warm air from the tropical region and cold air from the polar region. The deciduous forest biome has four distinct seasons: autumn, winter, spring, and summer. Every season has a significant change in weather. The average temperature is fairly mild. From summer to winter, though, the changes in temperature are significant. The annual average temperature is 50° Fahrenheit (10° Celsius). In summer, the average temperature is 70°F (21°C), and winter temperatures average a bit below 32°F (0°C). Overnight, temperatures do not drop that much. There is a moist cover of air that **insulates** the forest at night. During the summer, this layer keeps the forest cooler during the day.

Precipitation in this biome occurs throughout the year. Deciduous forests receive about 30 to 60 inches (75 to 153 centimeters) of precipitation annually. The precipitation contributes to the moisture of this forest and to bodies of water. These forests have rivers, streams, creeks, and springs. There are also lakes, ponds, and marshes that provide homes to various plants and animals.

Rivers and streams dispose of excess water. They help manage the precipitation the deciduous forest receives each year.

FASCINATING FACTS

Weather is hard to predict in temperate zones because it can change in a day, an hour, or a few minutes.

Deciduous forests have some of the best growing conditions because they have rich soil, moderate precipitation, and a long growing season.

The Photosynthesis Process

Sunlight

Oxygen

Carbon Dioxide

Water

Glucose

During photosynthesis, water and carbon dioxide enter the leaves. Sunlight turns these materials into oxygen and glucose.

Global Warming

There is evidence that deciduous trees have an effect on climate change, specifically **global warming**. Through a process called photosynthesis, deciduous trees absorb carbon dioxide and turn it into oxygen. Carbon dioxide is a greenhouse gas. It helps keep Earth warm. However, when there is too much carbon dioxide in the air, Earth can become too warm, harming the environment and all living things in it.

Deciduous trees may play a role in reducing, or at least regulating, the effects of global warming. As Earth becomes warmer, the growing season of the deciduous forest biome is extended. When the warmth of spring arrives earlier than normal, trees produce their leaves earlier as well. Likewise, when autumn arrives late, the trees stay green for a longer period. When this happens, the trees take more carbon dioxide from the air. This keeps Earth's temperature from rising too high.

Deciduous Forest Seasons

T he deciduous forest biome has cold winters, hot summers, cool autumns, and warm springs. Spring normally arrives in April and ends in June, when the warmer air of summer moves in. Summer continues through June and July, and ends in late August. Autumn begins in September, and winter arrives in December.

The seasons of a deciduous forest can be observed by the changes to the trees. In spring, tiny buds appear on the branches of trees. These buds eventually develop into leaves. Throughout spring and summer, the leaves of a deciduous tree are green. The color of a tree's leaves are determined by **pigments**. A pigment called chlorophyll causes leaves to be green. Besides chlorophyll, leaves also have pigments called carotenoid and anthocyanin. Carotenoid colors a leaf yellow, orange, or brown. Anthocyanin colors it red. Chlorophyll and carotenoid are always in leaf cells, but the chlorophyll covers the carotenoid during the spring and summer months. Anthocyanin is only produced in the fall, and only in certain trees.

Inside a Leaf

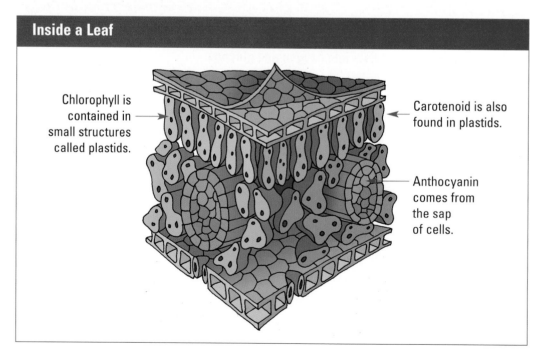

Chlorophyll is contained in small structures called plastids.

Carotenoid is also found in plastids.

Anthocyanin comes from the sap of cells.

A deciduous tree undergoes many changes over the course of a year.

Chlorophyll plays an important role in photosynthesis. It helps trees use sunlight to change water and carbon dioxide into food. Photosynthesis takes place during the warm spring and summer growing season.

As the temperature begins to cool and the days become shorter, trees slow their production of chlorophyll. The leaves begin changing color, becoming yellow, red, orange, and brown. This change signals the arrival of autumn.

The leaves are not strong enough to survive the freezing temperatures of winter. When this season arrives, the leaves dry up and fall to the ground. The trees are bare until spring, when the cycle begins anew.

FASCINATING FACTS

Autumn leaves can be used to identify some types of trees. The leaves of a birch tree are bright yellow, while a red maple's leaves will be bright scarlet.

Even though it is a deciduous tree, oak leaves do not fall from the tree in the winter.

Deciduous Forest Technology

It is important for people to study what affects forests, how animals and forests grow, and what can be done to keep forests alive. By studying plants and trees, scientists are able to modify and improve some qualities of trees. To do this, scientists study the genes of trees. Genes make up and determine certain characteristics of organisms. Certain genes can make a tree taller, shorter, a certain color, or more easily damaged by hail or insects. By studying the genes of trees, scientists can develop new, more robust, kinds of trees and plant them where they are needed. Gene research can also help develop trees for manufacturing wood products.

Scientists and biologists are developing **biotechnology** applications that help to protect forests and improve regrowth. This type of research is conducted in laboratories. Researchers determine what types of trees are genetically stronger and how trees' genetics differ. They also investigate how trees can be improved through **genetic engineering**, and study how to save forests. Genetic engineering can help new plants grow healthier, faster, and stronger.

By injecting genetic material into tree cells, scientists transfer the traits that improve tree growth. Scientists want to find new ways to make trees more tolerant to insects and animals that harm trees. This will ensure that forests survive for a long time.

Scientists use gene research to help trees survive insect infestations, including those of the spruce bark beetle in Alaska.

The FACE rings provide a way of experimenting with forests in their normal setting.

The relationship between deciduous forests and global warming is also of concern to scientists. Technologies such as Free Air Carbon Dioxide Enrichment (FACE) help scientists study how carbon dioxide levels impact deciduous trees. FACE technology allows scientists to study ecosystems within nature. The technology consists of a circle of pipes that are placed around a group of trees. Elevated levels of carbon dioxide are pumped from holding tanks to the rings, and out into the encircled trees. This simulates global warming conditions. Scientists then track the changes to the ecosystem within the rings. In this way, they can determine the role that deciduous forests play in helping or hindering global warming.

FASCINATING FACTS

More than 65,000 trees are used to produce one issue of the Sunday edition of the *New York Times*.

An average American uses about 100 feet (30.5 meters) of wood per year for newspapers, books, fences, boxes, and other products.

Almost 40,000 square miles (103,600 square km) of forest are cut down every year. Only about one-third of these trees are replanted.

LIFE IN DECIDUOUS FORESTS

The deciduous forest is home to plants and animals of all sizes. Trees tower over the forest floor, while smaller plants, such as shrubs and flowers, grow in their shadow. Birds, insects, and even some mammals use the trees as their home. To other animals, the trees provide protection. They hide within the thick forest to avoid predators.

Mushrooms are found growing along the forest floor and at the base of tree trunks.

PLANTS

Various plants grow in different layers of the forest. There are usually five layers of plants. Taller trees, such as oak, beech, elm, and maple, grow 60 to 100 feet (18 to 30 m). These trees form a cover over the forest. This makes it quite shaded, although sunlight does seep through the leaves. **Saplings** and other shorter trees compose the second layer. The third layer includes shrubs and bushes, such as hawthorn, dogwood, and holly, which grow around tall trees. Flowers, such as bluebells and primrose, along with berries and herbs, make up the fourth layer. The fifth layer consists of carpet moss, ferns, and milk-cap mushrooms, which cover the forest floor. Virginia creeper and bittersweet vines grow up tree trunks, climbing through several layers.

Dogwood trees grow in Mississippi's Holly Springs National Forest.

MAMMALS

The deciduous forest is home to large and small mammals. Squirrels and mice scurry across the forest floor, while bears, deer, and cougars roam freely. Some animals, such as bears, migrate or **hibernate** during winter. Other animals found within the deciduous forest include fox, skunk, and opossum.

Mountain lions can be found in both deciduous and coniferous forests.

Weaver ants build their nests in the leaves of deciduous trees.

BIRDS AND INSECTS

Different birds can be found throughout the deciduous forest biome during various seasons. Some stay year-round, while others **migrate** south in winter. Worms, beetles, ants, mites, centipedes, mosquitoes, and other insects live in deciduous forests. They play an important role **pollinating** buds of trees and other plants. This helps flowers and buds grow into berries and fruit. Insects also eat dead matter, such as rotting logs and dead leaves, on the forest floor. They break the dead matter into smaller pieces and turn it into soil.

Trees and Plants

Trees and Shrubs

Some of the best-known trees and shrubs are deciduous. Oak trees are known for their size and their longevity. They can live for as long as 400 years. Their trunks can grow to be as much as 32 feet (9 m) in circumference. Elm trees are shaped like the letter "V," narrow at the bottom and wide at the top. They line the streets of many cities and towns. The sugar maple is the source of many a sweet treat. Its sap can be made into maple sugar, maple syrup, and maple candy. The hawthorn shrub grows to about 13 feet (4 m) in height. It is known for its many medicinal qualities. The berries of the hawthorn are said to help people suffering from insomnia. Some people believe the berries may help prevent heart attacks.

Bluebells are not always blue. They can be violet, white, or even pink.

Flowers

The bluebell is commonly found in the deciduous forest. Known for its brilliant blue color and its bell-like shape, this flower often grows in clusters, creating a colorful carpet on the forest floor. Like the hawthorn shrub, the primrose also has medicinal uses. It, too, has been used to treat insomnia, and is considered by some to be a good remedy for headaches. The flowers of one type of primrose, the oxlip, are edible and are full of vitamins and minerals. The oxlip is sometimes used to treat a bad cough.

The oak tree symbolizes strength and endurance. It is the national tree of Germany, Great Britain, and the United States.

Early American Indians used the leaves of the lady fern to cover their food, especially drying berries.

Mosses and Ferns

Ferns and mosses grow throughout a deciduous forest. Mosses are normally found around the base of a tree, beside streams and rivers, or along the ground. Carpet moss is found along the ground. As it ages, it turns from a yellowish green to a deep, velvety green, forming a plush carpet along the forest floor. Lady ferns are known for their delicate, lacelike leaves. These ferns often group together in a circle. As new strands of fern grow outside of the circle, the inner strands die away. The circle then becomes a ring.

FASCINATING FACTS

Some dead leaves can rot in a few weeks on the floor of a deciduous forest. Oak leaves can stay preserved for over a year.

The soil in a deciduous forest is very fertile. This helps plants grow better. There are two types of soil in a deciduous forest: alfisols and ultisols. Alfisols soils are dark and rich. Ultisols have fewer nutrients and more clay.

The use of elm trees to line city streets has declined in recent years due to Dutch elm disease. The disease is caused by a fungus that the elm-bark beetle carries. Dutch elm disease spreads quickly and destroys the trees. Many towns have cut down their elms to eliminate the spread of the disease.

Mammals

Black Bears

Black bears are found throughout the world's deciduous forests. They are solitary animals that spend most of their days searching for food. While they are normally black in color, some black bears are actually chocolate or cinnamon brown. Black bears weigh between 100 and 300 pounds (45 and 136 kilograms), with some bears weighing even more. These bears are omnivores. They eat both animals and plants. Insects, nuts, berries, and deer fawns are just some of their food sources. During the winter months, most black bears hibernate. Depending on where they reside, hibernation can last up to seven months, and sometimes longer. The bears do not eat or drink during this time.

Only male deer grow antlers. The antlers are shed from January to March and grow out again in early spring.

White-tailed Deer

The white-tailed deer is another deciduous forest dweller. The coat of the white-tailed deer changes color with the seasons. In summer, the coat is tan or brown. In winter, it turns gray-brown. This helps camouflage the deer to avoid predators. The white-tailed deer is an herbivore. It eats only plants. In summer, it dines on the many green plants found in the forest. In fall and winter, it eats corn, nuts, and twigs. This deer can move quickly when it senses danger, reaching speeds of 40 miles (67 km) per hour. It can also jump heights of 9 feet (3 m) when fleeing a predator.

Black bears living along North America's northwest coast sometimes feed on spawning salmon.

The opossum is one of Earth's oldest surviving mammals. It has been around for at least 70 million years.

Opossums

The opossum is North America's only **marsupial**. It has a pointed nose and a naked tail, and is about the size of a large house cat. Opossums build nests in hollows in the ground or in fallen trees. Like the black bear, the opossum is an omnivore. Its diet includes insects, mice, fruit, nuts, and seeds. When threatened, the opossum will hiss, growl, and bare its teeth. It may also "play possum" by pretending to die. This involves rolling over and stiffening. An opossum can stay in this position up to four hours.

FASCINATING FACTS

A group of bears is sometimes called a sloth.

An opossum has a **prehensile** tail. It can be used to grasp objects. It can also be wrapped around a tree branch to hold the opossum in place. Sometimes, an opossum will use its tail to hang from a branch, but not for very long.

Deer are ruminants. This means their stomachs have four chambers. The first two chambers are used to break down food and turn it into cud. The cud is then brought up to the deer's mouth, where it is chewed one more time. When the cud is swallowed, it enters the third chamber, and all of the fluid is removed. The food then enters the fourth chamber, where it is sent to the small intestine to have all of the nutrients absorbed.

Badgers and chipmunks enter torpor during winter months. This is a sleepy state, but they can still wake up.

Birds and Insects

Migrating Birds

Many of the birds that live in deciduous forests migrate south when it gets cold. These birds live in South or Central America for most of the year. They fly to deciduous forests in spring, where they raise their families over the summer months. The vireo is one type of bird that migrates to the deciduous forest. These birds are known for their constant singing. It is estimated that different vireo species have between 12 and 100 songs in their repertoire. The cerulean warbler flies from South America to southeastern North America every year. It prefers to stay in the upper zones of the forest and can often be seen flitting through the treetops. The great crested flycatcher is a welcome site in North America's deciduous forests. A natural pest controller, this bird eats more than 50 kinds of beetles, as well as many other insects.

Flycatchers capture insects that are in flight. This is how the birds received their name.

The blue jay has a variety of calls and is able to mimic the calls of other birds.

Year-round Residents

Woodpeckers, blue jays, and nuthatches live in the deciduous forest year-round. During winter, they eat stored seeds, buds, and berries. Woodpeckers peck at tree trunks to make holes for their nests or to find insects to eat. The birds are specially designed for this type of work. Their head and neck muscles are strong and can withstand the constant pecking. As well, their stiff tail feathers help brace them upright on the tree. Many birds that live in deciduous forests do not build nests on branches. Wrens and nuthatches use abandoned woodpecker nests for their homes.

Insects

Ants, butterflies, bees, and beetles are some of the insects found in a deciduous forest. Bees and other insects play an important role in this biome by helping with plant pollination. Flowers attract the bees, butterflies, and moths with their colors, their shape, and their sweet nectars. Flies are attracted to the smell of the flowers of red trillium and skunk cabbage. All of these insects pollinate flowers, which helps the plants grow and develop seeds.

One bee will visit anywhere from 50 to 1,000 flowers in a day.

FASCINATING FACTS

Spittlebug nymphs, one of the deciduous forest's many insects, cover themselves with spit-like bubbles that help hide them. These bubbles also keep them moist when they are sucking plant juices.

Birds have different methods to find insects to eat. Woodpeckers search branches and trunks, flycatchers catch them in midair, and towhees dig in the dirt.

Flies usually lay their eggs on dead animals, but the scent of some flowers smell similar to these animals. This tricks flies into laying their eggs on the flowers.

Some animals only live in one forest layer, but others, such as birds, move through all layers.

Endangered Land and Animals

There are few original deciduous forests left in the world. Many forests in North America are second growth forests. These are forests that have been cleared, or cut down, but have grown back. Others have been planted by people in areas where there were previously no trees. China is home to several humanmade forests.

Between 1870 and 1970, the United States lost much of its original deciduous forest growth. Almost 500 million acres (200 million hectares) of forest in the eastern United States alone was clear-cut. Initially, the trees were removed to make way for agricultural land. The United States government encouraged settlers to work the land and build their homes in these forests. As the country developed, the trees were cut down to build homes, factories, and other buildings.

Removing trees impacts the many plants and animals that live in the deciduous forest. Animals rely on the forest for shelter and food. When the trees are removed, many of the other plants found in the forest cannot survive. The animals that feed on them lose their food source. Some are able to move to another area. Others, however, die of starvation. The passenger pigeon used to be found throughout North America's deciduous forests. Due partially to habitat loss, it became extinct in 1913. In order to save the animals of the deciduous forests, the government has now set aside land as parks and preserves. Some animal populations, such as wild turkeys and bears, are starting to increase because of this wildlife protection.

Forests once covered 48 percent of Earth's land surface. Now, they cover only half of this area.

Acid rain damages a tree's leaves, causing them to fall off.

Another threat to deciduous forests is air pollution and acid rain. Pollution weakens trees, making them more susceptible to insects, cold weather, and disease. Eventually, the trees die. Air pollution often comes in the form of acid rain and ground-level ozone. Both are created by industrial and transportation pollution, including car exhaust. Groups of trees in the southern Appalachians have died as a result of air pollution and acid rain.

FASCINATING FACTS

Deer populations can grow so much that the deer become a threat to themselves. Due to their high numbers, they exhaust their food sources. Many deer die of starvation.

Forests usually grow back faster if the land was logged or burned rather than cleared, plowed, and farmed.

There is more forest in New England now than 100 years ago because old farms have been abandoned and new forests have grown there.

DECIDUOUS FOREST STUDIES

Many jobs relating to the deciduous forest require a background in math, biology, and other sciences. In these jobs, people study plants and animals, and how they live and grow. Some positions determine how climate or pollution affects the environment. Others study why some plants grow in certain climates or areas.

FOREST ECOLOGIST

- Duties: studies environmental factors that affect forests

- Education: bachelor's, master's, or doctoral degree in science or forest science

- Interests: plants, biology, environment

Forest ecologists conduct studies to see why certain trees grow in different places. They study tree history, light, and soil requirements, and how resistant trees are to diseases. They see how different types of trees adapt to new environmental conditions including soil, climate, and altitude.

BOTANIST

- Duties: studies life processes of plants

- Education: bachelor's, master's, or doctoral degree in botany, biology, plant sciences, or genetics

- Interests: math, biology, chemistry, biology, living organisms, environment

Botanists study plants and plant cells. They use microscopes and other scientific equipment to study plant structures. They also determine the effect of rainfall, temperature, soil, and elevation on plant growth.

BIOLOGIST

- Duties: studies plant and animal life

- Education: bachelor's, master's, or doctoral degree in science or biology

- Interests: animals, plants, research, anatomy, and environment

Different types of biologists study different aspects of the environment. They can study the origin, relationship, and functions of plants and animals. They also determine the environmental effects of land use and inform governmental representatives about their findings.

ECO CHALLENGE

1 What makes leaves green?

2 Why are deciduous forests cleared?

3 How many layers of trees and plants grow in a deciduous forest?

4 How much precipitation does a deciduous forest receive in a year?

5 What does deciduous mean?

6 How many square miles (square km) of deciduous forest are cut down every year? How much of this is replanted?

7 Name three birds that live in the deciduous forest all year long.

8 How many trees are used in a single issue of the Sunday edition of the *New York Times*?

9 Why are insects important to the deciduous forest?

10 When did the first deciduous trees appear on Earth?

Answers

10. about 180 million years ago
9. They pollinate flowers, help make soil, and are food for animals.
8. 65,000
7. blue jay, nuthatch, woodpecker
6. 40,000 square miles (103,600 sq km); one-third
5. to fall or shed at a certain time
4. 30 to 60 inches (75 to 133 cm)
3. five
2. for agriculture, farms, and homes
1. chlorophyll

ACID IN YOUR RAIN

Acid rain falls in all parts of the world. This is because wind moves the clouds that contain rain. Even an area that does not have factories or automobiles can have acid rain. Try this experiment to see how much acid is in the rain that falls in your area.

MATERIALS

- a clear jar with a wide mouth
- a rainy day
- litmus paper with accompanying color chart

1. Remove the lid from the jar. Place the jar in an open area outdoors, where rain can fall directly into it.

2. Wait for a rainy day, and let the rain gather in the jar.

3. When the rain stops, pick up the jar, and take it indoors.

4. Dip a strip of litmus paper into the water.

5. Compare the color of the litmus paper to the color chart. Is your rain **acidic**, **alkaline**, or **pH** balanced?

FURTHER RESEARCH

How can I find out more information about ecosystems, deciduous forests, and wildlife?

- Libraries have many interesting books about ecosystems, deciduous forests, and wildlife.

- Science centers, museums, and interpretive programs at zoos and parks are great places to learn about ecosystems, deciduous forests, and wildlife.

- The Internet offers some great websites dedicated to ecosystems, deciduous forests, and wildlife.

BOOKS

Braun, Emma Lucy. *Deciduous Forests of Eastern North America.* Caldwell, NJ: The Blackburn Press, 2001.

Darke, Richard. *The American Woodland Garden: Capturing the Spirit of the Deciduous Forest.* Portland, OR: Timber Press, 2002.

Delcourt, Hazel R. *Forests in Peril: Tracking Deciduous Trees from Ice-Age Refuges into the Greenhouse World.* Granville, OH: Mcdonald and Woodward Publishing Company, 2002.

WEBSITES

Where can I learn more about the world's biomes?

worldbiomes.com
www.worldbiomes.com

Where can I learn more about the deciduous forest biome?

Blue Planet
www.blueplanetbiomes.org/deciduous_forest.htm

Where can I learn about animals that live in the deciduous forest?

Enchanted Learning
www.enchantedlearning.com/biomes/tempdecid/tempdecid.shtml

GLOSSARY

acidic: forming or yielding an acid

alkaline: a base that neutralizes an acid

biotechnology: the use of living organisms in manufacturing and environmental management

ecosystem: a community of living things sharing an environment

fertile: able to produce plant life

genetic engineering: scientific change of the genetic structure of a living organism

global warming: an increase in the average temperature of Earth's atmosphere; enough to cause climate change

hibernate: to pass winter in a resting state

insulates: protects from heat, electricity, or sound

marsupial: a mammal that bears its young in a pouch on the abdomen

migrate: to move from one place to another place by season

organisms: living things

pH: a scale, ranging from zero to fourteen, that is used to assess the acid or alkaline content of a substance

pigments: substances that add color

pollinating: carrying pollen to a plant so that it is fertilized

prehensile: adapted for grasping or taking hold of something

saplings: young trees

INDEX